my life cycle

Sea Turtle

CHERRY LAKE PRESS

Published in the United States of America by Cherry Lake Publishing Group
Ann Arbor, Michigan
www.cherrylakepublishing.com

Reading Adviser: Marla Conn, MS, Ed., Literacy specialist, Read-Ability, Inc.
Book Designer: Jennifer Wahl
Illustrator: Jeff Bane

Photo Credits: © ymgerman/Shutterstock.com, 5; © topten22photo/Shutterstock.com, 7; © Magdalena Paluchowiska/Shutterstock.com, 9; © mundosemfim/Shutterstock.com, 11; © Kjersti Joergensen/Shutterstock.com, 13; © Michael Smith ITWP/Shutterstock.com, 15; © Shane Myers Photography/Shutterstock.com, 17; © Tony Northrup/Shutterstock.com, 19; © David Evison/Shutterstock.com, 21; © Julian Wiskemann/Shutterstock.com, 2; Cover, 2-3, 6, 10, 22, 24, Jeff Bane

Cherry Lake Press is an imprint of Cherry Lake Publishing Group.

Library of Congress Cataloging-in-Publication Data

Names: Gray, Susan Heinrichs, author. | Bane, Jeff, 1957- illustrator.
Title: Sea turtle / Susan H. Gray ; illustrated by Jeff Bane.
Description: Ann Arbor, Michigan : Cherry Lake Publishing, 2021. | Series: My life cycle | Includes index. | Audience: Grades 2-3
Identifiers: LCCN 2020030603 (print) | LCCN 2020030604 (ebook) | ISBN 9781534180024 (hardcover) | ISBN 9781534181731 (paperback) | ISBN 9781534181038 (pdf) | ISBN 9781534182745 (ebook)
Subjects: LCSH: Sea turtles--Life cycles--Juvenile literature.
Classification: LCC QL666.C536 G73 2021 (print) | LCC QL666.C536 (ebook) | DDC 597.92/8156--dc23
LC record available at https://lccn.loc.gov/2020030603
LC ebook record available at https://lccn.loc.gov/2020030604

Printed in the United States of America
Corporate Graphics

table of contents

About the author: Susan H. Gray has a master's degree in zoology. She loves writing science books, especially about animals. Susan lives in Arkansas with her husband, Michael. They think turtles are peaceful, pleasant animals.

About the illustrator: Jeff Bane and his two business partners own a studio along the American River in Folsom, California, home of the 1849 Gold Rush. When Jeff's not sketching or illustrating for clients, he's either swimming or kayaking in the river to relax.

I'm a baby sea turtle. I am warm inside an egg. My mom buried me in the sand to keep me safe.

One day, I **hatch**. Baby turtles are everywhere. I have 100 brothers and sisters!

We work together. We dig our way up through the sand.

When we **emerge**, we can see the light on the ocean. We hurry toward it.

Right away, I discover I can swim! Sometimes, I just float or hide out in **seaweed**.

For years, I take it easy. I swim, eat, and grow.

One day, I find a **mate**.
I'll be laying eggs soon.

I return to the beach. I find the perfect spot in the sand. That's where I start digging a hole.

I lay 100 eggs in the hole and cover them with sand. Then, I go back to the ocean.

My **hatchlings** will appear soon. But for now, my babies are warm and **protected**.

glossary & index

glossary

emerge (uh-MURJ) to move out of something

hatch (HACH) to break out of an egg

hatchlings (HACH-lingz) young animals that have broken out of their eggs

mate (MATE) a partner to produce babies with

protected (pruh-TEK-tid) in a safe place

seaweed (SEE-weed) plants growing in the ocean

index